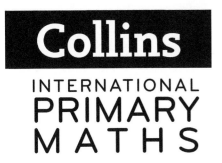

Collins

INTERNATIONAL PRIMARY MATHS

T0187288

Progress Book 6

William Collins' dream of knowledge for all began with the publication of his first book in 1819. A self-educated mill worker, he not only enriched millions of lives, but also founded a flourishing publishing house. Today, staying true to this spirit, Collins books are packed with inspiration, innovation and practical expertise. They place you at the centre of a world of possibility and give you exactly what you need to explore it.

Collins. Freedom to teach.

Published by Collins
An imprint of HarperCollins*Publishers*
The News Building
1 London Bridge Street
London
SE1 9GF

HarperCollins*Publishers*
1st Floor Watermarque Building
Ringsend Road
Dublin 4
Ireland

Browse the complete Collins catalogue at
www.collins.co.uk

British Library Cataloguing-in-Publication Data
A catalogue record for this publication is available from the British Library.

Author: Peter Clarke
Series editor: Peter Clarke
Publisher: Elaine Higgleton
Product developer: Holly Woolnough
Copyeditor: Tanya Solomons
Proofreader: Catherine Dakin
Answer checker: Steven Matchett
Cover designer: Gordon MacGilp
Cover illustrator: Ann Paganuzzi
Illustrator: Ann Paganuzzi
Typesetter: Ken Vail Graphic Design
Production controller: Lyndsey Rogers
Printed and Bound in the UK using 100% Renewable Electricity at CPI Group (UK) Ltd

Photo acknowledgements
Every effort has been made to trace copyright holders. Any omission will be rectified at the first opportunity.
p42c BlueRingMedia/Shutterstock; p42b Nicfz/Shutterstock.

MIX
Paper from
responsible sources
FSC
www.fsc.org **FSC™ C007454**

Contents

Introduction

The Progress Books include photocopiable end-of-unit progress tests which are designed to assist teachers with medium-term 'formative' assessment.

Each test is designed to be used within the classroom at the end of a Collins International Primary Maths unit to help measure the progress of learners and identify strengths and weaknesses.

Analysis of the results of the tests helps teachers provide feedback to individual learners on their specific strengths and areas that require improvement, as well as analyse the strengths and weaknesses of the class as a whole.

Self-assessment is also an important feature of the Progress Books as feedback should not only come from the teacher. The Progress Books provide opportunities at the end of each test for learners to self-assess their understanding of the unit, as well as space for teacher feedback.

Structure of the Progress Books

There is one progress test for each of the 27 units in Stage 6.

Each test consists of two pages of questions aimed at assessing the learning objectives from the Cambridge Primary Mathematics Curriculum Framework (0096) for the relevant unit. Where appropriate, this also includes questions to assess learners' development in one or more of the Thinking and Working Mathematically characteristics, as indicated by the TWM star on the page. All of the questions are typeset on triangles to indicate that they are suitable for the majority of learners and assess the unit's learning objectives.

Pages 5 to 9 include a list of photocopiable *I can* statements for each unit which are aimed at providing an opportunity for learners to undertake some form of self-assessment. The intention is that once learners have answered the two pages of questions, they turn to the *I can* statements for the relevant unit and think about each statement and how easy or hard they find the topic. For each statement they colour in the face that is closest to how they feel:

☺ I can do this   ☺ I'm getting there   ☹ I need some help.

A photocopiable variation of the Thinking and Working Mathematically Star is also included in the Progress Books. This version of the star includes *I can* statements for the eight TWM characteristics. Its purpose is to provide an opportunity for learners, twice a term/semester, to think about each of the statements and record how confident they feel about Thinking and Working Mathematically.

Administering each end-of-unit progress test

Recommended timing: 20 to 30 minutes, although this can be altered to suit the needs of individual learners and classes.

Before starting each end-of-unit progress test, ensure that each learner has the resources needed to complete the test. If needed, resources are listed in the 'You will need' box at the start of each test.

On completion of each end-of-unit progress test, use the answers and mark scheme available as a digital download to mark the tests.

Use the box at the bottom of the second page of each end-of-unit progress test to either:

- write the number of marks achieved by the learner out of the total marks possible.

- sign or initial your name to indicate you have marked the test.

- draw a simple picture or diagram such as one of the three faces (☺, ☺, ☹) to indicate your judgement on the learner's level of understanding of the unit's learning objectives.

- write a brief comment such as 'Well done!', 'You've got it', 'Getting there' or 'See me'.

Provide feedback to individual learners as necessary on their strengths and the areas that require improvement. Use the 'Class record-keeping document' located at the back of the Teacher's Guide and as a digital download to update your judgement of each learner's level of mastery in the relevant sub-strand.

I can statements

At the end of each unit, think about each of the *I can* statements and how easy or hard you find the topic. For each statement, colour in the face that is closest to how you feel.

Unit 1 – Counting and sequences	Date:			
• I can count on and count back in steps of constant size, including fractions and decimals, and extend beyond zero to include negative numbers.		☺	😐	☹
• I can find missing terms in a number sequence.		☺	😐	☹
• I can use the position of a term in a sequence to work out its value, including a sequence involving square numbers.		☺	😐	☹
Unit 2 – Addition and subtraction of whole numbers (A)	Date:			
• I can add a positive number to a negative number, such as $-7 + 9$.		☺	😐	☹
• I can solve missing number problems, such as $a + 8 = 16$ or $16 - a = 5$.		☺	😐	☹
Unit 3 – Addition and subtraction of whole numbers (B)	Date:			
• I can find the difference between positive and negative numbers.		☺	😐	☹
• I can find the difference between two negative numbers.		☺	😐	☹
• I can work out the value of variables in calculations.		☺	😐	☹
• I can use a simple formula for given values.		☺	😐	☹
Unit 4 – Multiples, factors, divisibility, squares and cubes	Date:			
• I understand and can find common multiples.		☺	😐	☹
• I understand and can find common factors.		☺	😐	☹
• I can recognise numbers that are divisible by 3, 6 and 9.		☺	😐	☹
• I can use square numbers to recognise cube numbers.		☺	😐	☹

Unit 5 – Whole number calculations	Date:			
• I know which property of number to use to simplify calculations.		☺	😐	☹
• I understand that the four operations follow a particular order.		☺	😐	☹
• I understand that when completing a calculation that includes brackets, operations in brackets must be completed first.		☺	😐	☹
Unit 6 – Multiplication of whole numbers	Date:			
• I can multiply numbers to 10 000 by a 1-digit number.		☺	😐	☹
• I can multiply numbers to 10 000 by a 2-digit number.		☺	😐	☹
• I can estimate the answer to a multiplication calculation.		☺	😐	☹
Unit 7 – Division of whole numbers (A)	Date:			
• I can divide numbers to 100 by a 1-digit number.		☺	😐	☹
• I can divide numbers to 1000 by a 1-digit number.		☺	😐	☹
• I can estimate the answer to a division calculation.		☺	😐	☹
Unit 8 – Division of whole numbers (B)	Date:			
• I can divide numbers to 100 by a 2-digit number.		☺	😐	☹
• I can divide numbers to 1000 by a 2-digit number.		☺	😐	☹
• I can estimate the answer to a division calculation.		☺	😐	☹
Unit 9 – Place value and ordering decimals	Date:			
• I can explain the value of the tenths, hundredths and thousandths digits in decimals.		☺	😐	☹
• I can compose and decompose decimals.		☺	😐	☹
• I can regroup decimals in different ways.		☺	😐	☹
• I can compare and order decimals.		☺	😐	☹
Unit 10 – Place value, ordering and rounding decimals	Date:			
• I can multiply whole numbers and decimals by 10, 100 and 1000.		☺	😐	☹
• I can divide whole numbers by 10, 100 and 1000.		☺	😐	☹
• I can divide decimals by 10 and 100.		☺	😐	☹
• I can round decimals to the nearest tenth and whole number.		☺	😐	☹

Unit 11 – Fractions (A)	Date:			
• I understand that finding a fraction can be thought of as a division of the numerator by the denominator.		☺	😐	☹
• I can simplify a fraction to its lowest terms.		☺	😐	☹
• I can compare and order fractions with different denominators.		☺	😐	☹
Unit 12 – Fractions (B)	Date:			
• I understand that fractions can act as operators.		☺	😐	☹
• I can add fractions with different denominators.		☺	😐	☹
• I can subtract fractions with different denominators.		☺	😐	☹
• I can multiply fractions by a whole number.		☺	😐	☹
• I can divide fractions by a whole number.		☺	😐	☹
Unit 13 – Percentages	Date:			
• I can recognise percentages of shapes.		☺	😐	☹
• I can calculate percentages of whole numbers and quantities.		☺	😐	☹
• I can compare and order percentages of quantities.		☺	😐	☹
Unit 14 – Addition and subtraction of decimals	Date:			
• I can add pairs of decimals mentally.		☺	😐	☹
• I can estimate and add pairs of decimals using a written method.		☺	😐	☹
• I can subtract pairs of decimals mentally.		☺	😐	☹
• I can estimate and subtract pairs of decimals using a written method.		☺	😐	☹
Unit 15 – Multiplication of decimals	Date:			
• I can estimate and multiply decimals by a 1-digit number.		☺	😐	☹
• I can estimate and multiply decimals by a 2-digit number.		☺	😐	☹
Unit 16 – Division of decimals	Date:			
• I can estimate and divide one-place decimals by a 1-digit number.		☺	😐	☹
• I can estimate and divide two-place decimals by a 1-digit number.		☺	😐	☹

Unit 17 – Proportion and ratio	Date:			
• I understand the relationship between two quantities when they are in direct proportion.		☺	😐	☹
• I can use equivalent ratios to calculate unknown amounts.		☺	😐	☹
Unit 18 – 2D shapes and symmetry	Date:			
• I can identify, describe, classify and sketch different quadrilaterals.		☺	😐	☹
• I can identify and label parts of a circle.		☺	😐	☹
• I can construct circles of a given radius or diameter.		☺	😐	☹
• I can identify rotational symmetry in familiar shapes, patterns or images.		☺	😐	☹
Unit 19 – 3D shapes	Date:			
• I can identify, describe and sketch compound 3D shapes.		☺	😐	☹
• I can identify and sketch different nets for different 3D shapes.		☺	😐	☹
Unit 20 – Angles	Date:			
• I can classify, estimate and use a protractor to measure angles.		☺	😐	☹
• I can use a protractor to draw angles.		☺	😐	☹
• I can calculate unknown angles in a triangle.		☺	😐	☹
Unit 21 – Measurements, including time	Date:			
• I understand the relationship between units of time, and can convert between them, including times expressed as a fraction or decimal.		☺	😐	☹
• I understand the difference between volume and capacity.		☺	😐	☹
Unit 22 – Area and surface area	Date:			
• I can prove that the area of a right-angled triangle is half the area of its related rectangle.		☺	😐	☹
• I can use the area of a related rectangle to find the area of a right-angled triangle.		☺	😐	☹
• I can understand the relationship between area of 2D shapes and surface area of 3D shapes.		☺	😐	☹

Unit 23 – Coordinates	Date:			
• I can read coordinates in all four quadrants.		☺	😐	☹
• I can plot coordinates in all four quadrants.		☺	😐	☹
• I can plot points to form shapes in all four quadrants.		☺	😐	☹
• I can plot points to form lines in all four quadrants.		☺	😐	☹
Unit 24 – Translation, reflection and rotation	Date:			
• I can translate a shape across all four quadrants on a coordinate grid.		☺	😐	☹
• I can reflect shapes in horizontal, vertical and diagonal mirror lines.		☺	😐	☹
• I can rotate shapes 90° around a vertex clockwise or anticlockwise.		☺	😐	☹
Unit 25 – Statistics (A)	Date:			
• I can record, organise, represent and interpret data in a Venn and Carroll diagram to answer a question.		☺	😐	☹
• I can record, organise, represent and interpret data in a tally chart, frequency table and bar chart to answer a question.		☺	😐	☹
• I can record, organise, represent and interpret data in a waffle diagram and pie chart to answer a question.		☺	😐	☹
• I can find and interpret the mode, median, mean and range of a data set.		☺	😐	☹
Unit 26 – Statistics (B)	Date:			
• I can record, organise, represent and interpret data in a frequency table and frequency diagram to answer a question.		☺	😐	☹
• I can record, organise, represent and interpret data in a line graph and dot plot to answer a question.		☺	😐	☹
• I can record, organise, represent and interpret data in a scatter graph to answer a question.		☺	😐	☹
Unit 27 – Probability	Date:			
• I can use the language of proportion and probability to predict, describe and compare the probability of different events.		☺	😐	☹
• I can identify two events that are mutually exclusive or independent.		☺	😐	☹

Number

Name: _____

1 Count on or back in the steps given.

a Count on in steps of 0·3.

6·25, 6·55, 6·85, 7·15, ☐, ☐, ☐, 8·35, ☐, ☐

b Count back in steps of 0·04.

12·6, 12·56, 12·52, ☐, 12·44, ☐, ☐, 12·32, ☐, ☐

c Count on in steps of $\frac{2}{3}$.

$1\frac{1}{3}$, 2, $2\frac{2}{3}$, ☐, ☐, $4\frac{2}{3}$, ☐, ☐, ☐

d Count back in steps of $\frac{2}{5}$.

$9\frac{4}{5}$, $9\frac{2}{5}$, 9, ☐, ☐, ☐, $7\frac{2}{5}$, ☐, ☐

e Count back in steps of 6.

15, 9, 3, ☐, ☐, ☐, −21, ☐, −33, ☐

f Count on in steps of 0·2.

−1·8, −1·6, ☐, −1·2, ☐, ☐, −0·6, ☐, ☐,

g Count back in steps of 0·05.

0·1, 0·05, 0, ☐, ☐, −0·15, ☐, ☐, ☐

h Count on in steps of $\frac{1}{4}$.

$-1\frac{1}{2}$, $-1\frac{1}{4}$, −1, ☐, ☐, $-\frac{1}{4}$, ☐, ☐, ☐

2 Write the missing numbers in the table and answer the questions.

a

Position	1	2	3	4	5	6	7	8
Term	6	12	18	24				

b What is the position-to-term rule? _____

Number

c What is the value of the term in the 12th position? ☐

d What is position of the term with a value of 120? ☐

3 9, 18, 27, 36 are the first four terms in a sequence.

② **a** What is the position-to-term rule? _____

b What is the value of the term in the 8th position? ☐

c What is the position of the term with a value of 108? ☐

4 Complete the table for square numbers up to 10^2.

Position	Calculation	Notation	Value
1	1 × 1	1^2	1
9			
6			
3			
8			
5			
2	2 × 2	2^2	4
10			
4			
7			

5 Use the values in **4** to complete the sequence of square numbers.

1, 4, ☐, 16, ☐, ☐, ☐, 64, ☐, ☐,

Now look at and think about each of the *I can* statements. ☐

Date: _____

Number

Name: _____

1 Use the number line to find each answer.

−20 −19 −18 −17 −16 −15 −14 −13 −12 −11 −10 −9 −8 −7 −6 −5 −4 −3 −2 −1 0 1 2 3 4 5 6 7 8 9 10 11 12 13 14 15 16 17 18 19 20

a −12 + 7 = ☐

b −15 + 21 = ☐

c −14 + 12 = ☐

d −16 + 24 = ☐

e −18 + 17 = ☐

f −6 + 17 = ☐

g −11 + 11 = ☐

h −8 + 15 = ☐

i −10 + 2 = ☐

j −9 + 19 = ☐

2 Use the thermometer to find each answer.

a The temperature increases from −2 °C by 8 °C.

What is the new temperature? ☐

b The temperature is −8 °C and rises by 4 °C.

What is the new temperature? ☐

c What is 12 degrees more than −5 °C? ☐

d At 6:00 a.m. the temperature is −10 °C.

By 8:00 a.m. the temperature has risen 6 °C.

By 10.00 a.m. is has risen another 8 °C.

What is the temperature at 10.00 a.m.? ☐

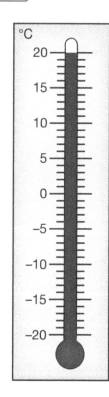

3 Complete the calculations.

a −37 + 26 = ☐

b −26 + 19 = ☐

c −38 + 33 = ☐

d −29 + 51 = ☐

e −45 + 27 = ☐

f −36 + 39 = ☐

Number

g −31 + 68 = ☐

h −48 + 25 = ☐

i −41 + 33 = ☐

j −52 + 57 = ☐

 4 Work out the unknown values.

a $54 + a = 80$

$a = $ ☐

b $b + 55 = 99$

$b = $ ☐

c $38 + c = 57$

$c = $ ☐

d $92 = 34 + d$

$d = $ ☐

e $e + 27 = 71$

$e = $ ☐

f $14 + f = 76$

$f = $ ☐

g $101 = g + 48$

$g = $ ☐

h $85 = h + 52$

$h = $ ☐

i $18 + i = 84$

$i = $ ☐

j $89 = j + 11$

$j = $ ☐

k $k + 41 = 72$

$k = $ ☐

k $25 + l = 93$

$l = $ ☐

 5 Work out the unknown values.

a $a − 43 = 43$

$a = $ ☐

b $77 − b = 34$

$b = $ ☐

c $74 − c = 18$

$c = $ ☐

d $42 = d − 43$

$d = $ ☐

e $e − 79 = 16$

$e = $ ☐

f $9 = 66 − f$

$f = $ ☐

g $29 = g − 26$

$g = $ ☐

h $h − 40 = 49$

$h = $ ☐

i $44 = 76 − i$

$i = $ ☐

j $34 − j = 19$

$j = $ ☐

k $k − 17 = 61$

$k = $ ☐

l $84 − l = 55$

$l = $ ☐

Now look at and think about each of the *I can* statements. ☐

Date: _____

Number

Name: _____

1 Use the number line to find the difference between each pair of numbers.

–20 –19 –18 –17 –16 –15 –14 –13 –12 –11 –10 –9 –8 –7 –6 –5 –4 –3 –2 –1 0 1 2 3 4 5 6 7 8 9 10 11 12 13 14 15 16 17 18 19 20

a –15 and 9 ☐

b 6 and –12 ☐

c –11 and 14 ☐

d –18 and 16 ☐

e 15 and –7 ☐

f –17 and 17 ☐

g –10 and 15 ☐

h 2 and –6 ☐

i –18 and 8 ☐

j –9 and 2 ☐

2 Use the number line to find the difference between each pair of numbers.

–20 –19 –18 –17 –16 –15 –14 –13 –12 –11 –10 –9 –8 –7 –6 –5 –4 –3 –2 –1 0 1 2 3 4 5 6 7 8 9 10 11 12 13 14 15 16 17 18 19 20

a –13 and –4 ☐

b –6 and –10 ☐

c –17 and –1 ☐

d –10 and –5 ☐

e –18 and –3 ☐

f –20 and –10 ☐

g –2 and –14 ☐

h –20 and –2 ☐

i –19 and –15 ☐

j –15 and –7 ☐

3 Answer each problem.

a A lift descends from a basement car park at level –1 to level –6.

How many floors has the lift descended? ☐

b A diver is swimming at a depth of –4 metres. He descends to a depth of –9 metres.

How many more metres has the diver descended? ☐

c The temperature in the morning is –9°C. By the evening, it has fallen to –17°C.

By how many degrees has the temperature fallen? ☐

Number

 4 $a + b = 10$ and a and b are whole numbers. What are all the possible solutions for the values of a and b? Complete the table of solutions.

a											
b											

 5 $c + d = 20$ and c and d are both whole numbers between 3 and 17. c and d are also both even numbers. What are all the possible solutions for the values of c and d? Complete the table of solutions.

c						
d						

6 Write an equation using the letters e and f that describes the relationship between the two values below. Then find all the possible solutions.

Assume all numbers are whole numbers.

The sum of the two numbers is 18.

Both numbers are odd numbers between 2 and 16.

a Equation:

b Solutions:

 7 $g - h = 15$

Use the equation to complete the table.

g	20		18		41		53
h		27		50		54	

Now look at and think about each of the *I can* statements.

Date: _____

Number

Name: _____

 List the first 15 multiples of 3 and 4. Then write the common multiples of 3 and 4.

a Multiples of 3 [   ]

b Multiples of 4 [   ]

c Common multiples of 3 and 4 [   ]

 Write the factors in the correct part of each Venn diagram.

a
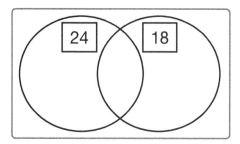

b
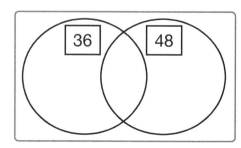

 Circle all the numbers that are divisible by 3.

178	381	923	739	801	455
849	671	510	625	566	264

4 How do you know a number is divisible by 3?

②

_____

_____

 Circle all the numbers that are divisible by 6.

286	354	649	906	413	864
594	260	838	532	192	741

6 How do you know a number is divisible by 6?

②

_____

_____

Circle all the numbers that are divisible by 9.

324	592	936	748	893	675
185	441	623	286	336	504

 How do you know a number is divisible by 9?

⟨2⟩

_____

_____

 Complete the table for cube numbers up to 53.

Notation	Calculation	Value
1^3	$1 \times 1 \times 1$	1
2^3		
3^3		
4^3		
5^3		

 Use your knowledge of square numbers to complete each calculation.
The first one has been done for you.

$2^3 = \boxed{2^2} \times \boxed{2}$

$ = \boxed{4} \times \boxed{2}$

$ = \boxed{8}$

a $5^3 = \boxed{} \times \boxed{}$

$ = \boxed{} \times \boxed{}$

$ = \boxed{}$

b $4^3 = \boxed{} \times \boxed{}$

$ = \boxed{} \times \boxed{}$

$ = \boxed{}$

c $3^3 = \boxed{} \times \boxed{}$

$ = \boxed{} \times \boxed{}$

$ = \boxed{}$

Now look at and think about each of the *I can* statements.

Date: _____

Number

Name: _____

1 Solve each calculation mentally.

a $3 + 6 \times 5 =$ ☐

b $19 + 22 + 21 + 38 =$ ☐

c $2 \times 4 \times 3 \times 5 =$ ☐

d $9 - 2 \times 4 =$ ☐

e $3 \times 6 \times 2 + 8 =$ ☐

f $57 + 28 + 43 =$ ☐

2 Write a calculation for each problem, then simplify and solve it.
Show any working out.

a A theatre has three floors. On each floor there are 26 rows of seats with 12 seats in each row. What is the total number of seats in the theatre?

☐ $=$ ☐

b Marta has 8 parcels to post. The masses her parcels are 530 g, 940 g, 750 g, 800 g, 670 g, 250 g, 560 g and 400 g. What is the total mass of the parcels?

☐ $=$ ☐

c A restaurant has 20 tables. At 16 of the tables, there are 4 people sitting at each table. At the other tables there are 2, 5, 6 and 8 people. How many people is this altogether?

☐ $=$ ☐

3 Simplify and solve each calculation. Show any working out.

a $(14 + 8) \times 3 =$ ☐

b $63 \div (9 - 2) =$ ☐

Number

c $(28 - 14) \times 15 =$ ▢

d $16 \times (8 + 7) =$ ▢

e $(38 + 28) \div 3 =$ ▢

f $4 \times (56 - 18) =$ ▢

4 Write a calculation for each problem, then simplify and solve it.

Show any working out.

a Five days a week Sunita travels 17 km to and from work. Each Saturday she travels a total of 12 km to visit her parents. How far does Sunita travel each week?

▢ $=$ ▢

b Mika buys 8 trays of seedlings. Each tray has 24 seedlings. He also buys another tray of 36 seedings. Altogether how many seedlings does Mika buy?

▢ $=$ ▢

c A baker cooks muffins in trays of 12 and cupcakes in trays of 10. She cooks 8 trays of muffins and 15 trays of cupcakes. How many muffins and cupcakes does she cook altogether?

▢ $=$ ▢

Now look at and think about each of the *I can* statements.

▢

Date: _____

Number

Name: _____

1 Estimate, then use the **expanded written method** to solve each calculation. Show your working out.

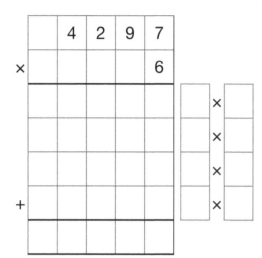

a Estimate: [＿＿＿]

		4	2	9	7
×					6

b Estimate: [＿＿＿]

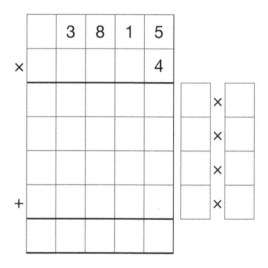

		3	8	1	5
×					4

2 Estimate, then use the **formal written method** to solve each calculation. Show your working out.

a Estimate: [＿＿＿]

		4	5	7
×				8

b Estimate: [＿＿＿]

		6	2	7
×				4

c Estimate: [＿＿＿]

		5	7	2	6
×					7

d Estimate: [＿＿＿]

		4	9	8	3
×					6

e Estimate: [＿＿＿]

		3	8	6	4
×					8

f Estimate: [＿＿＿]

		2	7	0	6
×					9

3 Estimate, then use the **grid method** to solve each calculation. Show your working out.

a 547 × 42 = [ ]

Estimate: [ ]

× [ ] [ ] [ ]

b 1393 × 28 = [ ]

Estimate: [ ]

× [ ] [ ] [ ] [ ]

4 Estimate, then use the **formal written method** to solve each calculation. Show your working out.

a Estimate: [ ]

			2	7	4
×				3	8

b Estimate: [ ]

			3	6	4	2
×					4	7

Now look at and think about each of the *I can* statements.

[ ]

Date: _____

Number

Name: _____

1 Estimate, then use the **expanded written method** to solve each calculation. Show your working out.

a 82 ÷ 3 = [   ]

Estimate: [   ]

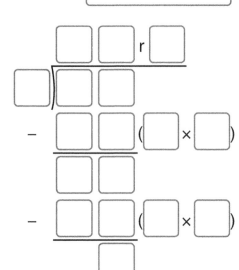

b 98 ÷ 6 = [   ]

Estimate: [   ]

2 Estimate, then use **short division** to solve each calculation. Show your working out.

a 76 ÷ 3 = [   ]

Estimate: [   ]

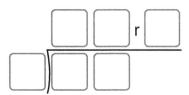

b 91 ÷ 4 = [   ]

Estimate: [   ]

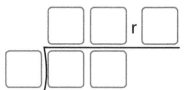

c 93 ÷ 5 = [   ]

Estimate: [   ]

d 87 ÷ 2 = [   ]

Estimate: [   ]

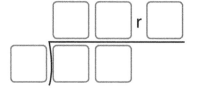

3 Estimate, then use the **expanded written method** to solve each calculation. Show your working out.

a $546 \div 8 =$ [ ]

Estimate: [ ]

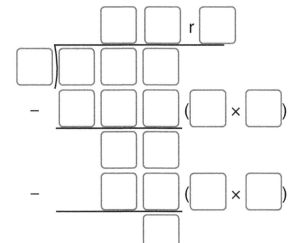

b $642 \div 7 =$ [ ]

Estimate: [ ]

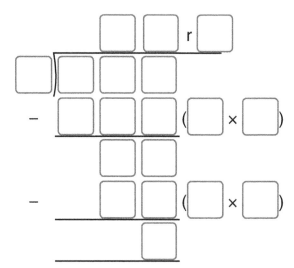

4 Estimate, then use **short division** to solve each calculation. Show your working out.

a $274 \div 5 =$ [ ]

Estimate: [ ]

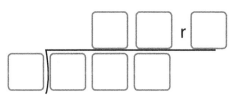

b $638 \div 7 =$ [ ]

Estimate: [ ]

c $531 \div 8 =$ [ ]

Estimate: [ ]

d $452 \div 6 =$ [ ]

Estimate: [ ]

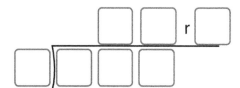

Now look at and think about each of the *I can* statements. [ ]

Date: _____

Number

Name: _____

1 Estimate, then use the **expanded written method** to solve each calculation. Show your working out.

a 86 ÷ 12 = [ ]

Estimate: [ ]

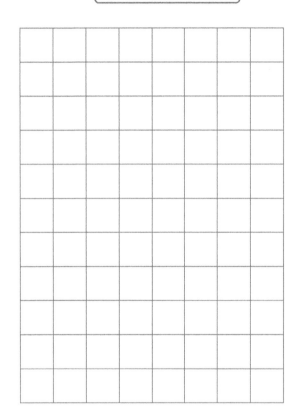

b 95 ÷ 14 = [ ]

Estimate: [ ]

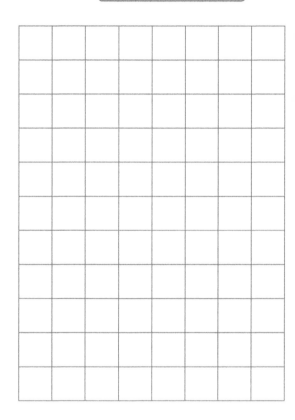

2 Estimate, then use the compact form of the **expanded written method** to solve each calculation. Show your working out.

a 93 ÷ 13 = [ ]

Estimate: [ ]

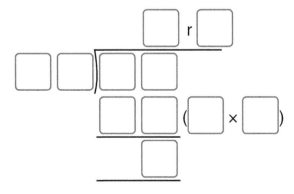

b 84 ÷ 15 = [ ]

Estimate: [ ]

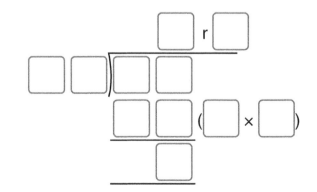

3 Estimate, then use the **expanded written method** to solve each calculation. Show your working out.

a 888 ÷ 24 = [ ]

Estimate: [ ]

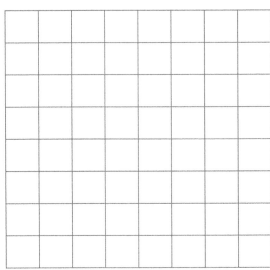

b 576 ÷ 32 = [ ]

Estimate: [ ]

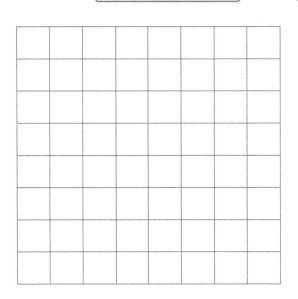

 3 Estimate, then use **long division** to solve each calculation. Show your working out.

a 392 ÷ 14 = [ ]

Estimate: [ ]

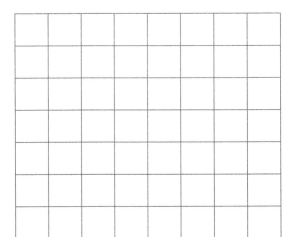

b 494 ÷ 26 = [ ]

Estimate: [ ]

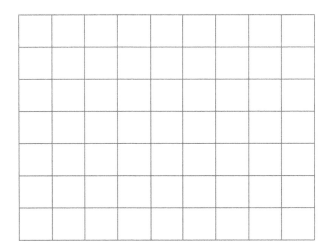

Now look at and think about each of the *I can* statements. [ ]

Date: _____

Name: _____

Number

 1 Write the next number in each sequence.

Count on in steps of...						
a 0·03	0·134	0·164	0·194			
b 0·2	1·268	1·468	1·668			
c 0·004	−0·284	−0·28	−0·276			

2 Write the decimal that is equivalent to each fraction.

a $\frac{15}{100}$ = [ ]

b $\frac{8}{1000}$ = [ ]

c $\frac{62}{1000}$ = [ ]

d $\frac{7}{100}$ = [ ]

e $\frac{6}{10}$ = [ ]

f $\frac{534}{1000}$ = [ ]

3 Decompose the numbers by place value.

a 0·841 = [ ] + [ ] + [ ]

b 2·438 = [ ] + [ ] + [ ] + [ ]

c 0·405 = [ ] + [ ]

4 Complete each statement.

a 6·387 is composed from [ ]

b 2·054 is composed from [ ]

c 34·82 is composed from [ ]

 5 Write the decimal.

a 3 ones, 7 tenths, 8 hundredths and 4 thousandths [ ]

b 8 tenths, 9 ones, 7 thousandths and 6 hundredths [ ]

c 2 hundredths, 4 ones, 1 thousandth and 3 tenths [ ]

d 4 thousandths, 7 hundredths, 5 tenths and 8 ones [ ]

Number

6 Decompose each number in four different ways.

a

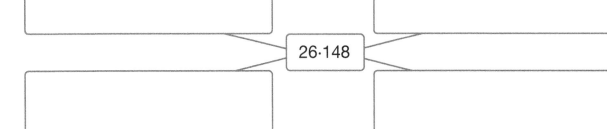

3·675

b

26·148

7 Write the correct symbol, < or >, to compare each pair of decimals.

a 3·25 ☐ 3·27

b 4·18 ☐ 4·2

c 11·33 ☐ 11·43

d 24·04 ☐ 24·03

e 35·61 ☐ 35·29

f 45·92 ☐ 45·29

8 Order each set of decimals.

a 3·42, 3·33, 3·23, 3·43, 3·24   ☐ < ☐ < ☐ < ☐ < ☐

b 7·53, 7·59, 7·54, 7·52, 7·58   ☐ < ☐ < ☐ < ☐ < ☐

c 9·02, 9·13, 9·23, 9·03, 9·2   ☐ < ☐ < ☐ < ☐ < ☐

d 5·5, 5·55, 5·05, 5, 5·51   ☐ < ☐ < ☐ < ☐ < ☐

Now look at and think about each of the *I can* statements.

☐

Date: _____

Number

Name: _____

1 Complete each calculation.

a $73 \times 10 =$ ☐

b $94 \div 100 =$ ☐

c $730 \div 10 =$ ☐

d $8 \times 1000 =$ ☐

e $568 \times 10 =$ ☐

f $210 \div 1000 =$ ☐

g $9 \times 100 =$ ☐

h $7 \div 10 =$ ☐

i $1649 \div 100 =$ ☐

j $67 \times 100 =$ ☐

k $961 \times 100 =$ ☐

l $812 \div 1000 =$ ☐

2 Complete each calculation.

a $583 \times$ ☐ $= 5830$

b $12 \div$ ☐ $= 0.12$

c $2400 \div$ ☐ $= 24$

d $146 \times$ ☐ $= 146\,000$

e $442 \div$ ☐ $= 44.2$

f $816 \div$ ☐ $= 0.816$

g $73 \times$ ☐ $= 73\,000$

h $94 \times$ ☐ $= 9400$

3 Complete each calculation.

a $62.5 \times 10 =$ ☐

b $8.3 \times 100 =$ ☐

c $37.6 \times 1000 =$ ☐

d $5.29 \times 10 =$ ☐

e $17.74 \times 1000 =$ ☐

f $40.29 \times 100 =$ ☐

g $3.21 \times 100 =$ ☐

h $583.9 \times 1000 =$ ☐

Number

 Complete each calculation.

a 20·6 ÷ 10 = ☐

b 4·7 ÷ 100 = ☐

c 136·7 ÷ 10 = ☐

d 0·2 ÷ 10 = ☐

e 26·1 ÷ 100 = ☐

f 105·9 ÷ 100 = ☐

g 306·2 ÷ 10 = ☐

h 55·3 ÷ 100 = ☐

 Complete each calculation.

a 7·5 ÷ ☐ = 0·075

b 2·56 × ☐ = 25·6

c 11·48 × ☐ = 1148

d 3·7 ÷ ☐ = 0·37

e 39·2 ÷ ☐ = 0·392

f 16·45 ÷ ☐ = 1·645

g 0·639 × ☐ = 63·9

h 21·55 × ☐ = 21 550

 Write the two one-place decimals that each number comes between. Then circle the number that the decimal rounds to.

a ☐ 13·27 ☐

b ☐ 9·53 ☐

c ☐ 24·38 ☐

d ☐ 11·24 ☐

 Write the two whole numbers that each decimal comes between. Then circle the number that the decimal rounds to.

a ☐ 18·64 ☐

b ☐ 2·48 ☐

c ☐ 37·16 ☐

d ☐ 12·87 ☐

Now look at and think about each of the *I can* statements.

Date: _____

Name: _____

Number

1 Write each fraction as a division.

a $\frac{1}{7}$ = ☐ ÷ ☐

b $\frac{9}{10}$ = ☐ ÷ ☐

c $\frac{8}{15}$ = ☐ ÷ ☐

2 Write each division as a fraction.

a $5 \div 8$ = $\frac{☐}{☐}$

b $2 \div 7$ = $\frac{☐}{☐}$

c $13 \div 10$ $\frac{☐}{☐}$

3 Write the fraction described and then express the fraction as a division.

a There are 12 vehicles in a car park. Five of them are motorbikes.

Number of motorbikes as a fraction of the total number of vehicles:

Fraction as a division: ☐ ÷ ☐

$\frac{☐}{☐}$

b There are 26 people visiting a lake. Nine of them are swimming.

Number of people swimming as a fraction of the total number of people:

Fraction as a division: ☐ ÷ ☐

$\frac{☐}{☐}$

4 Use the diagram to simplify each fraction.

a $\frac{5}{35}$ = $\frac{☐}{☐}$

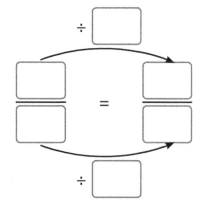

b $\frac{16}{20}$ = $\frac{☐}{☐}$

Number

5 Draw lines to match each fraction to its simplest form.

$\frac{42}{60}$ $\qquad$ $\frac{28}{48}$ $\qquad$ $\frac{18}{24}$ $\qquad$ $\frac{9}{27}$ $\qquad$ $\frac{40}{64}$ $\qquad$ $\frac{12}{15}$

$\frac{1}{3}$ $\qquad$ $\frac{7}{10}$ $\qquad$ $\frac{4}{5}$ $\qquad$ $\frac{3}{4}$ $\qquad$ $\frac{7}{12}$ $\qquad$ $\frac{5}{8}$

6 Use the diagram to convert one fraction to have the same denominator as the other. Then compare the fractions using < or >.

a $\frac{1}{4}$ [ ] $\frac{5}{12}$

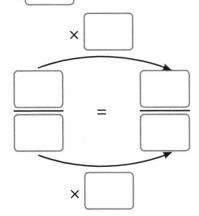

b $\frac{7}{10}$ [ ] $\frac{3}{5}$

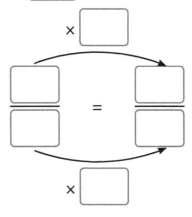

7 Write the correct symbol, < or >, to compare each pair of fractions.

a $\frac{2}{5}$ [ ] $\frac{3}{10}$ $\qquad$ **b** $\frac{3}{4}$ [ ] $\frac{1}{2}$ $\qquad$ **c** $\frac{5}{9}$ [ ] $\frac{2}{3}$

d $\frac{5}{6}$ [ ] $\frac{2}{3}$ $\qquad$ **e** $\frac{3}{4}$ [ ] $\frac{7}{8}$ $\qquad$ **f** $\frac{9}{12}$ [ ] $\frac{5}{6}$

8 Convert the fractions to the same denominator then order them.

a

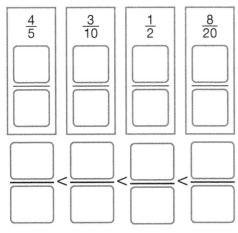

b

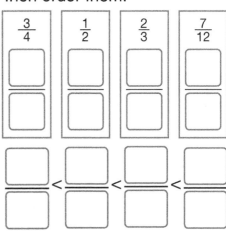

Now look at and think about each of the *I can* statements.

Date: _____

Number

Name: _____

 Work out these fractions. Show your working out.

a $\frac{5}{6}$ of $480 = \boxed{}$

b $\frac{3}{8}$ of 560g = $\boxed{}$

c $\frac{5}{9}$ of 126 = $\boxed{}$

d $\frac{8}{5}$ of 75km = $\boxed{}$

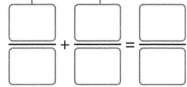 Solve each of the fraction calculations. Write answers greater than 1 as an improper fraction and a mixed number.

a $\frac{2}{3} + \frac{5}{9} = \frac{\boxed{}}{\boxed{}}$ or $\boxed{} \frac{\boxed{}}{\boxed{}}$

Common denominator: $\boxed{}$

$\frac{\boxed{}}{\boxed{}} + \frac{\boxed{}}{\boxed{}} = \frac{\boxed{}}{\boxed{}}$

b $\frac{2}{5} + \frac{3}{4} = \frac{\boxed{}}{\boxed{}}$ or $\boxed{} \frac{\boxed{}}{\boxed{}}$

Common denominator: $\boxed{}$

$\frac{\boxed{}}{\boxed{}} + \frac{\boxed{}}{\boxed{}} = \frac{\boxed{}}{\boxed{}}$

c $\frac{3}{4} - \frac{7}{10} = \frac{\boxed{}}{\boxed{}}$

Common denominator: $\boxed{}$

$\frac{\boxed{}}{\boxed{}} - \frac{\boxed{}}{\boxed{}} = \frac{\boxed{}}{\boxed{}}$

d $\frac{4}{5} - \frac{3}{7} = \frac{\boxed{}}{\boxed{}}$

Common denominator: $\boxed{}$

$\frac{\boxed{}}{\boxed{}} - \frac{\boxed{}}{\boxed{}} = \frac{\boxed{}}{\boxed{}}$

3 Use the diagrams to multiply. Write your answers as an improper fraction and a mixed number.

a $\frac{3}{4} \times 5 = \dfrac{\square}{\square}$ or $\square \dfrac{\square}{\square}$

 + + + +

b $\frac{7}{10} \times 3 = \dfrac{\square}{\square}$ or $\square \dfrac{\square}{\square}$

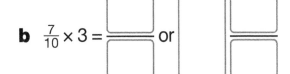

c $\frac{3}{5} \times 4 = \dfrac{\square}{\square}$ or $\square \dfrac{\square}{\square}$

d $\frac{5}{6} \times 5 = \dfrac{\square}{\square}$ or $\square \dfrac{\square}{\square}$

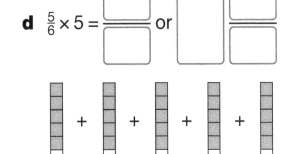

4 Draw your own area models to divide the fractions.

a $\frac{3}{4} \div 4 = \dfrac{\square}{\square}$

b $\frac{5}{8} \div 6 = \dfrac{\square}{\square}$

c $\frac{3}{4} \div 7 = \dfrac{\square}{\square}$

c $\frac{5}{6} \div 3 = \dfrac{\square}{\square}$

Now look at and think about each of the *I can* statements.

Date: _____

Name: _____

Number

1 The shaded part of each 100 grid represents a percentage. Write the percentage shown.

a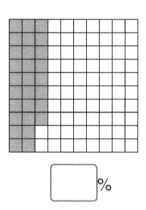

b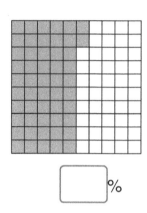

c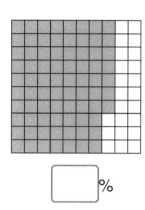

[ ] %   [ ] %   [ ] %

2 Shade each grid to show the percentage given.

a 64%   **b** 32%   **c** 96%

3 Complete the missing fractions, percentages and decimals.

a $\frac{35}{100}$ = [ ] % = [ ]

b $\frac{[\ ]}{[\ ]}$ = [ ] % = 0·3

c $\frac{[\ ]}{[\ ]}$ = 85% = [ ]

d $\frac{[\ ]}{[\ ]}$ = [ ] % = 0·15

e $\frac{8}{10}$ = [ ] % = [ ]

f $\frac{[\ ]}{[\ ]}$ = 45% = [ ]

4 Find the percentage of each quantity. Show your working out.

a 25% of $36 = [ ]

b 70% of 90 g = [ ]

Number

c 45% of 80 *l* = ☐

d 15% of 120 km = ☐

 5 Calculate the new price after the price increase or decrease.
Show your working out.

a 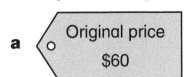 Original price $60    Price increase of 20%   

b  Original price $150    Price decrease of 75%   

 6 Write the correct symbol, < or >, to compare each statement.

a 20% ☐ $\frac{4}{10}$

b $\frac{9}{10}$ ☐ 80%

c 0·8 ☐ $\frac{3}{4}$

d 25% ☐ 0·3

e 0·9 ☐ 75%

f $\frac{3}{4}$ ☐ 0·6

7 Order the amounts. Write the letter codes in order of value, from smallest to greatest.

a

A	B	C
45% of 680 m*l*	$\frac{3}{10}$ of 900 m*l*	0·65 of 440 m*l*

☐ < ☐ < ☐

b

A	B	C
0·15 of $860	$\frac{2}{5}$ of $450	35% of $480

☐ < ☐ < ☐

Now look at and think about each of the *I can* statements. ☐

Date: _____

Unit 14 Addition and subtraction of decimals

Name: ____________________

1 Solve these calculations **mentally** using your preferred strategies.

Show your working out.

a 3·8 + 5·4 =

b 9·6 + 2·5 =

c 17·6 + 23·8 =

d 9·6 + 8·32 =

e 0·85 + 0·7 =

f 2·74 + 0·5 =

g 3·5 + 0·28 =

h 2·04 + 2·4 =

i 1·43 + 7·2 =

2 Estimate, then solve each calculation using the **expanded written method**. Show your working out.

a 58·73 + 4·678 =

Estimate:

			·			
+			·			
			·			
			·			
			·			
			·			
+			·			
			·			

b 8·663 + 47·85 =

Estimate:

			·			
+			·			
			·			
			·			
			·			
			·			
+			·			
			·			

Number

3 Estimate, then solve each calculation using the **formal written method**. Show your working out.

a 65·48 + 7·286 =

Estimate:

			.			
+			.			
			.			

b 9·604 + 28·29 =

Estimate:

			.			
+			.			
			.			

4 Solve these calculations **mentally** using your preferred strategies. Show your working out.

a 8·5 − 4·7 =

b 6·8 − 3·54 =

c 7·65 − 5·8 =

d 12·2 − 8·5 =

e 11·5 − 8·26 =

f 17·35 − 9·7 =

5 Estimate, then solve each calculation using the **formal written method**. Show your working out.

a 86·35 − 58·6 =

Estimate:

			.			
−			.			
			.			

b 32·431 − 6·85 =

Estimate:

			.			
−			.			
			.			

Now look at and think about each of the *I can* statements.

Date: _____

Name: _____

1 Estimate, then use the **grid method** to work out the answer to each calculation. Show your working out.

a 54·3 × 8 = [   ]

Estimate: [   ]

× [   ][   ][   ]

[   ][   ]

b 35·26 × 3 = [   ]

Estimate: [   ]

× [   ][   ][   ]

[   ][   ]

2 Estimate, then use the **expanded written method** to work out the answer to each calculation. Show your working out.

a 64·3 × 8 = [   ] × 8 ÷ 10

Estimate: [   ]

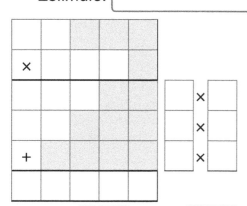

× [   ]
× [   ]
× [   ]

Answer: [   ] ÷ 10 = [   ]

b 49·85 × 3 = [   ] × 3 ÷ 100

Estimate: [   ]

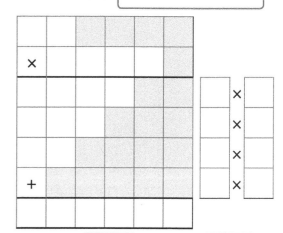

× [   ]
× [   ]
× [   ]
× [   ]

Answer: [   ] ÷ 100 = [   ]

3 Estimate, then use the **formal written method** to work out the answer to each calculation. Show your working out.

a 58·9 × 6 = [   ] × 6 ÷ 10

Estimate: [   ]

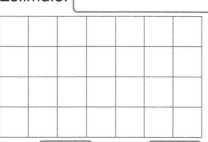

Answer: [   ] ÷ 10 = [   ]

b 73·52 × 4 = [   ] × 4 ÷ 100

Estimate: [   ]

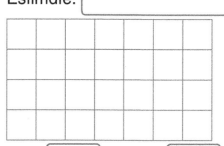

Answer: [   ] ÷ 100 = [   ]

Number

4 Estimate, then use the **grid method** to work out the answer to each calculation. Show your working out.

a 36·7 × 46 = [   ]

Estimate: [   ]

× [ ] [ ] [ ]

+ _____

_____

b 75·28 × 39 = [   ]

Estimate: [   ]

× [ ] [ ] [ ] [ ]

+ _____

_____

5 Estimate, then use the most appropriate mental or written method to calculate the answer to each calculation. Show your working out.

a 28·43 × 27 = [   ]

Estimate: [   ]

b 49·37 × 34 = [   ]

Estimate: [   ]

Now look at and think about each of the *I can* statements.

[ ]

Date: _____

Number

Name: _____

1 Estimate, then use the **expanded written method** to work out the answer to each calculation. Show your working out.

a 72·3 ÷ 3 = [ ]

Estimate: [ ]

72·3 ÷ 3 is equivalent to

[ ]

[ ] ÷ 10 = [ ]

b 145·2 ÷ 4 = [ ]

Estimate: [ ]

145·2 ÷ 4 is equivalent to

[ ]

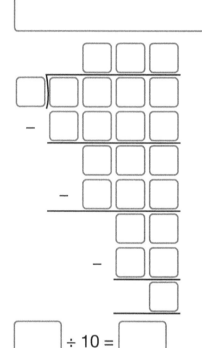

[ ] ÷ 10 = [ ]

2 Estimate, then use the **short division method** to work out the answer to each calculation. Show your working out.

a 86·8 ÷ 4 = [ ]

Estimate: [ ]

86·8 ÷ 4 is equivalent to

[ ]

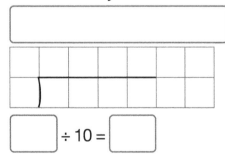

[ ] ÷ 10 = [ ]

b 274·2 ÷ 6 = [ ]

Estimate: [ ]

274·2 ÷ 6 is equivalent to

[ ]

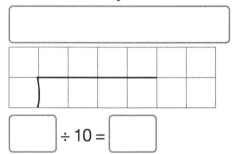

[ ] ÷ 10 = [ ]

Number

 3 Estimate, then use the **expanded written method** to work out the answer to each calculation. Show your working out.

a 48·96 ÷ 6 = [ ]

Estimate: [ ]

48·96 ÷ 6 is equivalent to

[ ]

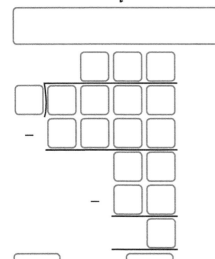

[ ] ÷ 100 = [ ]

b 36·84 ÷ 4 = [ ]

Estimate: [ ]

36·84 ÷ 4 is equivalent to

[ ]

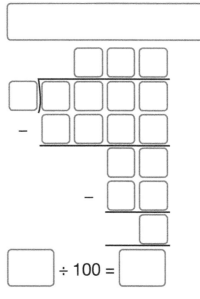

[ ] ÷ 100 = [ ]

 4 Estimate, then use the **short division method** to work out the answer to each calculation· Show your working out.

a 72·32 ÷ 8 = [ ]

Estimate: [ ]

72·32 ÷ 8 is equivalent to

[ ]

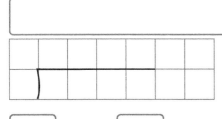

[ ] ÷ 100 = [ ]

b 84·91 ÷ 7 = [ ]

Estimate: [ ]

84·91 ÷ 7 is equivalent to

[ ]

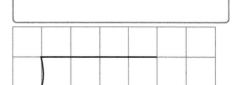

[ ] ÷ 100 = [ ]

Now look at and think about each of the *I can* statements.

[ ]

Date: _____

Number

Name: _____

1 Answer these questions. Show your working out.

a Two photos, photo A and photo B, are in proportion. Photo B is 3 times wider than photo A. If photo B is 30 cm wide, how wide is photo A?

b Two boxes, box A and box B, are in proportion. Box B is 5 times taller than box A. If box A is 12 cm tall, how tall is box B?

2 Answer these questions. Show your working out.

a How much do 4 cans of pop cost?

> 1 can for 80 c **POP**

b What is the cost of 20 tins of beans?

> 4 tins of beans for $2

c Leroy spends $24 on crisps. How many bags of crisps does he buy?

> 10 bags of crisps for $6

3 Use the recipe to work out the amounts.

a How much macaroni is needed for 2 people?

b How much flour is needed for 8 people?

c How much cheese is needed for 6 people?

d How many people can you serve with 3 litres of milk?

> **Macaroni and cheese**
> (Serves 4 people)
> 400 g macaroni
> 100 g butter
> 100 g flour
> 1 litre milk
> 200 g cheese
> 1 tablespoons oil

Number

4 Answer these questions. Show your working out.

   a The ratio of staff to guests at a campsite is 1:8.
      If there are 10 staff, how many guests are there?

   b The ratio of adults to children at a campsite is 2:5.
      If there are 35 children, how many adults are there?

   c The Singh family spent 14 days at the campsite on holiday.
      The ratio of sunny days to rainy days was 5:2.
      For how many days was it sunny?

5 Complete the table to help you answer these equivalent ratio problems.

For every 5 yellow flowers in a park there are 3 pink flowers.

   a If there are 35 yellow flowers,
      how many pink flowers are there?

   b If there are 18 pink flowers,
      how many yellow flowers are there?

   c If there are 60 yellow flowers,
      how many pink flowers are there?

   d If there are 42 pink flowers,
      how many yellow flowers are there?

Yellow flowers	5				
Pink flowers	3				

Now look at and think about each of the *I can* statements.

Date: _____

Geometry and Measure

Name: _____

You will need
- compass
- ruler

1 In the first row of the table, write the letter of each shape. Then complete the table by placing a tick ✓ or a cross ✗ in each box. You may need to write a short explanation in some of the boxes.

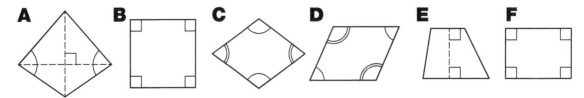

A   B   C   D   E   F

	parallelogram	rectangle	rhombus	square	trapezium	kite
letter						
opposite sides parallel						
opposite sides equal						
adjacent sides are equal						
diagonals are of equal length						
opposite angles equal						
four right angles						

2 Draw a line from each label to the correct part of the circle.

centre

circumference

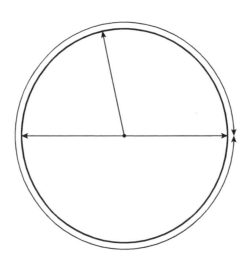

diameter

radius

Geometry and Measure

3 a Draw a circle with a **radius** of 3 cm.

b What is the diameter of the circle?

4 a Draw a circle with a **diameter** of 5 cm.

b What is the radius of the circle?

5 Write the order of rotational symmetry for each shape.

a

b

c

d
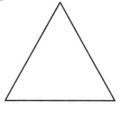

e
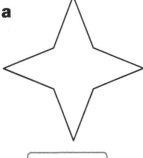

f
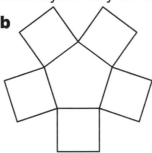

g

h
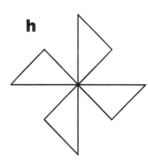

Now look at and think about each of the *I can* statements.

Date: _____

Name: _____

1 Each compound shape below is made of two component shapes. Name the component shapes.

A

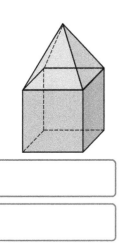

B

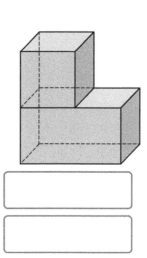

C

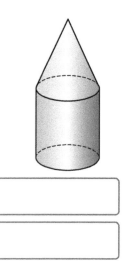

2 Look at the shapes in **1** and complete the table.

	Number of faces	Number of vertices	Number of edges
Shape A			
Shape B			
Shape C			

3 Copy the sketch of each shape. Try to include edges that are hidden from view.

a

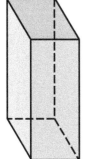

b

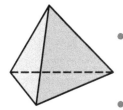

4 Write the letter that identifies each of the nets.

Letter	Net of a ...
A	cube
B	cuboid
C	triangular prism
D	pentaganal prism
E	square-based pyramid
F	pentagonal pyramid
G	cone
H	cylinder

5 Sketch the net of a triangular-based pyramid.

Now look at and think about each of the *I can* statements.

Date: _____

Geometry and Measure

Name: _____

You will need
• protractor
• ruler

1 Estimate the size of each angle, then measure it.

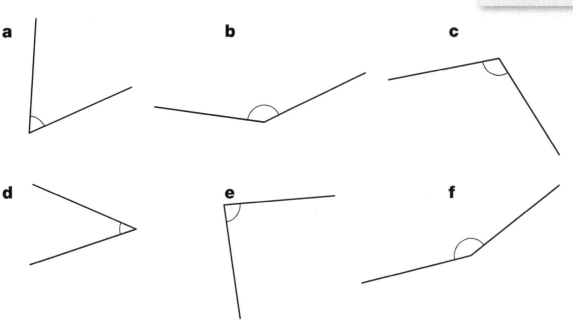

a   b   c

d   e   f

	a	b	c	d	e	f
Estimate						
Measurement						

2 Use a protractor to draw each of the angles.

a 39°

b 98°

c 53°

d 141°

48

Geometry and Measure

3 Calculate the missing angles. Show your working out.

a

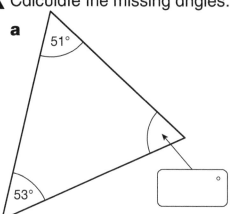

51°

53°

◯ °

b

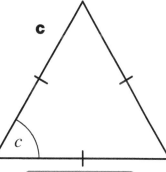

◻ °

45°

98°

c

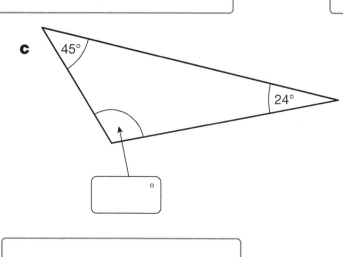

45°

24°

◻ °

d

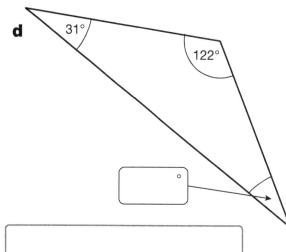

31°

122°

◻ °

4 Use your knowledge of triangles to calculate the missing angles.
Show your working out.

a

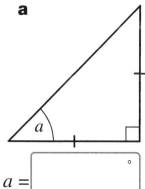

a

$a =$ ◻ °

b

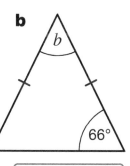

b

66°

$b =$ ◻ °

c

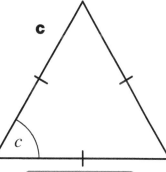

c

$c =$ ◻ °

d

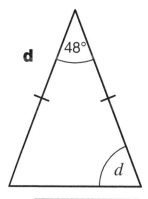

48°

d

$d =$ ◻ °

Now look at and think about each of the *I can* statements.

Date: _____

Name: _____

1 Convert the minutes to minutes and seconds. Show your working out.

a $2\frac{1}{2}$ min = [ ] min [ ] s

[ ]

b $4\frac{3}{4}$ min = [ ] min [ ] s

[ ]

c 3·5 min = [ ] min [ ] s

[ ]

d 1·25 min = [ ] min [ ] s

[ ]

e 2·7 min = [ ] min [ ] s

[ ]

f 6·2 min = [ ] min [ ] s

[ ]

2 Convert the hour to hours and minutes. Show your working out.

a $3\frac{1}{4}$ h = [ ] h [ ] min

[ ]

b $2\frac{3}{4}$ h = [ ] h [ ] min

[ ]

c 4·5 h = [ ] h [ ] min

[ ]

d 2·75 h = [ ] h [ ] min

[ ]

e $\frac{36}{10}$ h = [ ] h [ ] min

[ ]

f $\frac{94}{10}$ h = [ ] h [ ] min

[ ]

3 Write the capacity of each jug and the volume of water it contains.

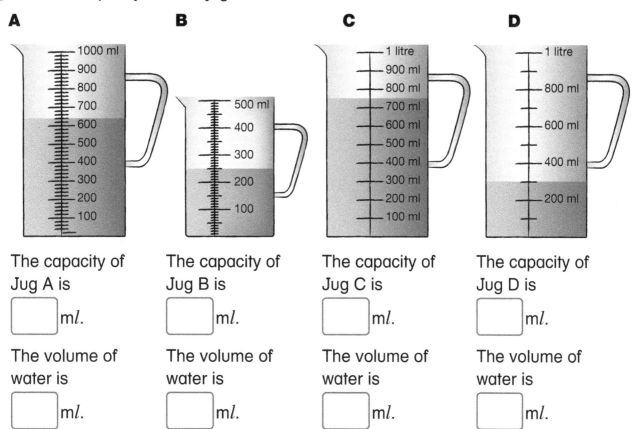

The capacity of
Jug A is

☐ ml.

The capacity of
Jug B is

☐ ml.

The capacity of
Jug C is

☐ ml.

The capacity of
Jug D is

☐ ml.

The volume of
water is

☐ ml.

The volume of
water is

☐ ml.

The volume of
water is

☐ ml.

The volume of
water is

☐ ml.

4 Use the information in **3** to answer
these questions.
Show your working out

a   What is the total capacity
of Jugs A and D?

☐

b   What is the total volume
of water in Jugs A and D?

☐

c   What is the total capacity
of Jugs B and C?

☐

d   What is the total volume
of water in Jugs B and C?

☐

Now look at and think
about each of the
I can statements.

☐

Date: _____

Geometry and Measure

Geometry and Measure

Name: _____

1 Draw two rectangles on the squared paper with different dimensions but that both have an area of 20 square units. Mark each rectangle to create a triangle with an area of 10 square units.

2 Work out the area of each triangle by drawing the related rectangle.

A: = ☐ square units   B: = ☐ square units   C: = ☐ square units

3 Calculate the area of each triangle. Show your working out.

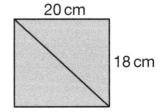

a 12 cm, 7 cm

Area of each triangle

= ☐ cm²

b 20 cm, 18 cm

Area of each triangle

= ☐ cm²

4 What are the different 2D shapes that cover the surface of each 3D shape? How many of each 2D shape are there?

a

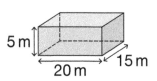

☐ _____

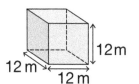

☐ _____

c

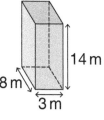

☐ _____

☐ _____

5 For each net, use a different coloured pencil to shade faces with the same area.

a

5m

15m

20m

b

12m

12m

12m

c

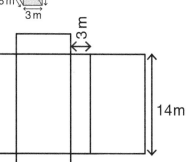

8m

Now look at and think about each of the *I can* statements.

☐

Date: _____

Geometry and Measure

Name: _____

You will need
• ruler

1 Write the coordinates for each letter.

Letter	Coordinates
A	
B	
C	
D	
E	
F	
G	
H	

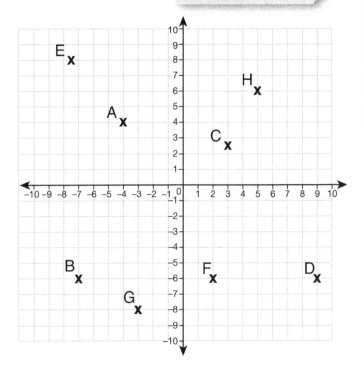

2 Plot points J to Q on the coordinate grid.

Letter	Coordinates
J	$(-6, -5)$
K	$(-5, 3)$
L	$(8\frac{1}{2}, 4)$
M	$(7, -9)$
N	$(-4, -7)$
O	$(9, 2)$
P	$(-2, 6.5)$
Q	$(4, -3\frac{1}{2})$

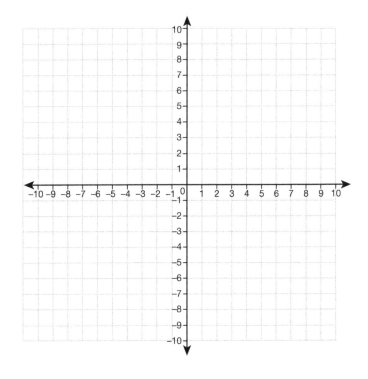

3 Draw each shape and identify the coordinates of the missing vertex.

a ABC is an isosceles triangle. The vertices of the base of the triangle are A (–9, –8), B (1, –8).
The triangle has a height of 10 units.
What are the coordinates of vertex C?

(_____, _____)

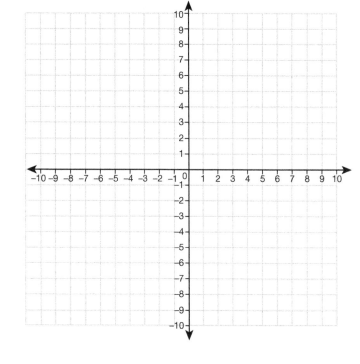

b DEFG is a parallelogram. D: (0, –3) E: (9, –3) F: (6, 8)

What are the coordinates of vertex G? (_____, _____)

4 Each pair of points lie on a different line. Plot and join the points with a straight line that extends to the edges of the coordinate grid. Complete the missing coordinates.

a H: (–7, 8) J: (2, –10)

Point K: (_____, 0)

Point L: (0, _____)

b M: (–7, –9) N: (2, 9)

Point P: (_____, 0)

Point Q: (0, _____)

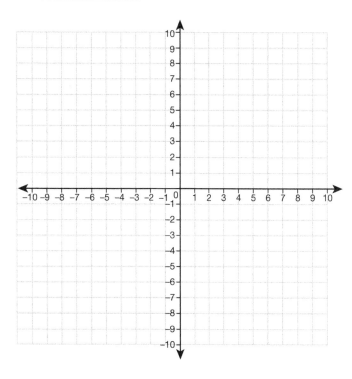

Now look at and think about each of the *I can* statements.

Date: _____

Name:
_____

You will need
• ruler

Geometry and Measure

 a Translate shape T: 5 units right, 3 units down. Label the translated shape T'.

b Write the coordinates of shape T'.

A' (____, ____)

B' (____, ____)

C' (____, ____)

D' (____, ____)

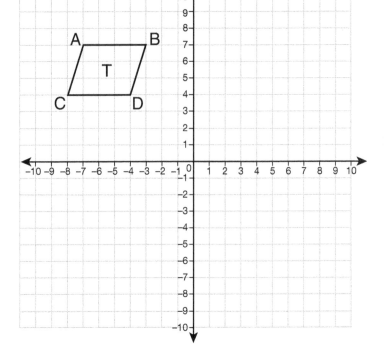

c Translate shape T: 10 units right, 13 units down. Label the translated shape T".

d Write the coordinates of shape T".

A" (____, ____) B" (____, ____) C" (____, ____) D" (____, ____)

2 Reflect each shape in the mirror line.

a

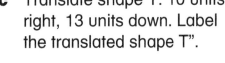

b

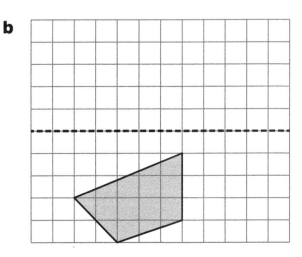

3 Reflect each shape in the mirror line.

a
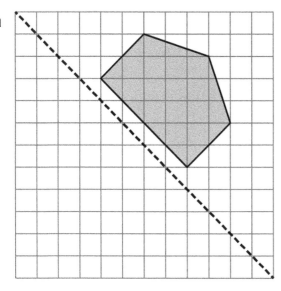

b
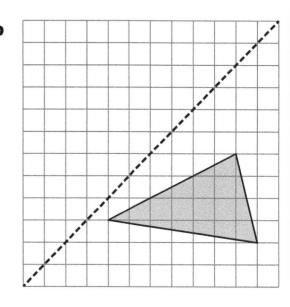

4 Draw the image of the triangle under a rotation of 90° anticlockwise about vertex A.

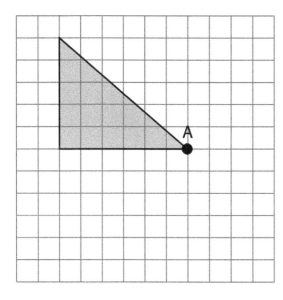

5 Draw the image of the trapezium under a rotation of 90° clockwise about vertex A.

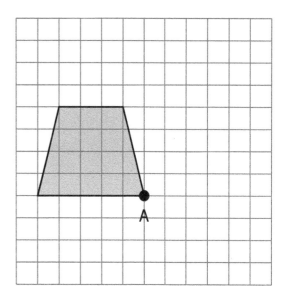

Now look at and think about each of the *I can* statements.

Date: _____

Statistics and Probability

Name: _____

You will need
- ruler
- coloured pencils

 1 Write three statements about the data in the Venn diagram.

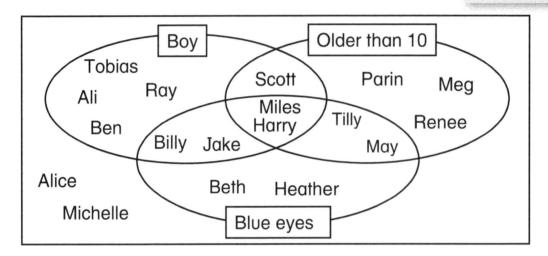

Boy | Older than 10

Tobias
Ali   Ray
Ben
Scott
Miles
Harry   Tilly
Parin   Meg
Renee
Billy   Jake   May
Alice
Michelle   Beth   Heather
Blue eyes

i. _____

ii. _____

iii. _____

2 Write the missing information in the tally chart. Then use the data to draw a bar chart. Don't forget to add a scale to the vertical axes

Favourite types of holiday

Holiday	Tally	Frequency
beach		11
skiing		8
adventure	ЦН|	
city break	ЦН| ЦН| ||||	

Number

3 Write three conclusions you can draw from the data in the bar chart.

i. _____

ii. _____

iii. _____

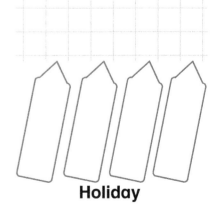

Holiday

4 Complete the frequency table. Then represent the data in a waffle diagram and a pie chart. You will need to complete the key.

Favourite TV programmes

Programme	Frequency	Percentage
comedy	4	
drama	2	
documentary	1	
sport	3	

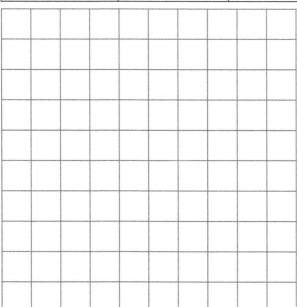

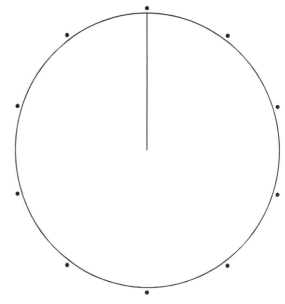

Key

5 Write three conclusions you can draw from the data in **4**.

i. _____

ii. _____

iii. _____

6 Find the mode, median, mean and range of the data set.

Computer game scores

12, 25, 16, 13, 24, 13, 15, 23, 11, 18, 22, 12, 16, 24, 13, 15

Mode: [ ]   Median: [ ]   Mean: [ ]   Range: [ ]

Now look at and think about each of the *I can* statements.

Date: _____

Statistics and Probability

Name: _____

 1 The owner of a gym took a quick survey to find out the average age of people in the gym. She recorded the information in a table. Draw a frequency diagram to represent the data. Remember to label the axes.

Age of gym attendees

Age	Frequency
20–30	12
30–40	10
40–50	6
50–60	8
60–70	2

2 Write three conclusions you can draw from the data in **1**.

i. _____

_____

_____

ii. _____

iii. _____

 3 Look at the frequency table in **1**. Use this data to draw a dot plot to display the data. Remember to label the horizontal axis.

4 Use the data in the table to complete the line graph.

⑥

Temperature on 26.11.2020

Time	Temp (°C)
00:00	−2
04:00	−4
08:00	0
12:00	5
16:00	8
20:00	2
00:00	−2

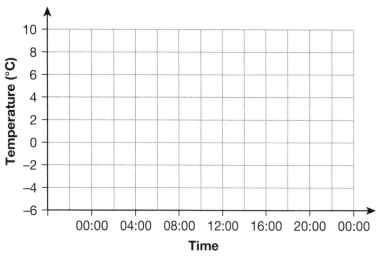

5 The data in the scatter graph represents the relationship between ice cream sales and change in temperature.

④

a Draw a line of best fit on the scatter graph.

b Write two conclusions you can draw from the data in the scatter graph.

i. _____

_____

_____

_____

_____

ii. _____

_____

_____

_____

_____

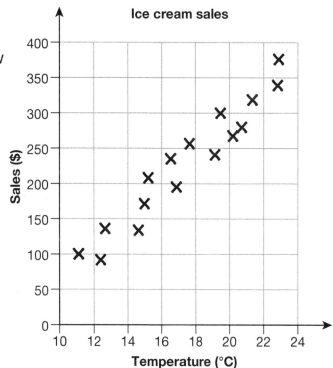

Now look at and think about each of the *I can* statements.

Date: _____

Name: _____

Statistics and Probability

1 Give your answers as a proportion and a percentage. What is the predicted probability of spinning:

a an even number? ⬚ in ⬚ or ⬚ %

b a number less than 8?

⬚ in ⬚ or ⬚ %

c a number greater than 8?

⬚ in ⬚ or ⬚ %

d a number between 2 and 7?

⬚ in ⬚ or ⬚ %

e a multiple of 3?

⬚ in ⬚ or ⬚ %

2 Joel conducts an experiment and spins the spinner in **1** 100 times and records the outcomes.
What do the experimental results show? Compare them to the predicted probabilities.

Number spun (outcome)	Frequency
1	9
2	10
3	12
4	10
5	7
6	13
7	8
8	11
9	10
10	10

a Probability of spinning an even number?

_____

_____

b Probability of spinning a number less than 8?

_____

c Probability of spinning a multiple of 5?

_____

3 Referring to the spinner in **1**, compare the probabilities of the events using the symbols >, < or =.

a an even number ⬚ an odd number

b a number more than 6 ⬚ a number less than 6

c a multiple of 2 ⬚ a multiple of 3

d a number from 1 to 4 ⬚ a number from 5 to 10

Statistics and Probability

4 Write a pair of events that have independent outcomes involving flipping a coin and rolling a 1–6 dice.

_____

_____

5 Write a pair of events that have mutually exclusive outcomes involving picking a card at random from a set of 0–20 number cards.

_____

_____

6 Joel spins a spinner. He spins it 200 times and records the number each time. Answer the questions. Show your working out.

a How many times would you expect Joel to spin a circle?

b How many times would you expect Joel to spin a square?

c How many times would you expect Joel to spin a rectangle?

d How many times would you expect Joel to spin a square or a triangle?

e How many times would you expect Joel to spin a shape that is not a rectangle?

Now look at and think about each of the *I can* statements.

Date: _____

The Thinking and Working Mathematically Star

Think about each of these *I can* statements and record how confident you feel about **Thinking and Working Mathematically**.

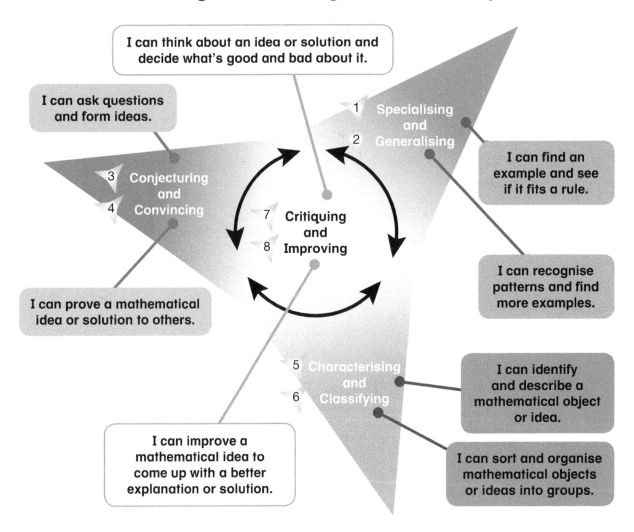

How confident do you feel Thinking and Working Mathematically?

Term ❶

Date: _____

☺ 😐 ☹

Date: _____

☺ 😐 ☹

Term ❷

Date: _____

☺ 😐 ☹

Date: _____

☺ 😐 ☹

Term ❸

Date: _____

☺ 😐 ☹

Date: _____

☺ 😐 ☹